Skittles

Contents

The Little Monkeys

Two little monkeys swinging in a tree,
Along came another one and then there were three.

Three little monkeys, look how fast they climb,
Along came another six and then there were nine.

Nine little monkeys doing silly tricks,
Three fell down and then there were six.

Six little monkeys swing and swing and then,
Along came another four and that made ten.

$6+4=10$

Ten little monkeys didn't hear the lion,
He grabbed one of them and then there were nine.

$10-1=9$

Nine little monkeys swinging in the sun,
Eight ran away and then there was one.

$9-8=1$

3

Swimming

Two happy children run off to swim,
Along come another two and four go splashing in.

Four laughing children shout, "The water's cool",
Along come another two and six are in the pool.

Six giggling children jump in and out,
Along come another two, now eight splash about.

6+2=?   8+2=?

Eight splashing children bob their heads below,
Two more jump in, now ten are in a row.

Ten happy children line up for a dive,
But only half are brave enough, so that left five.

Five shaky children wishing they would try,
"Come on" called the others, "it isn't very high."

10-5=?

Humpty-Dumpties

One humpty-dumpty sitting on the wall.
One humpty-dumpty sitting on the wall.
If one humpty-dumpty to three more should call,
There'd be four humpty-dumpties sitting on the wall.

Four humpty-dumpties sitting on the wall.
Four humpty-dumpties sitting on the wall.
And if four humpty-dumpties to three more should call,
There'd be seven humpty-dumpties sitting on the wall.

Seven humpty-dumpties sitting on the wall.
Seven humpty-dumpties sitting on the wall.
And if seven humpty-dumpties to three more should call,
There'd be ten humpty-dumpties sitting on the wall.

Ten humpty-dumpties sitting on the wall.
Ten humpty-dumpties sitting on the wall.
And if nine humpty-dumpties should topple off the wall,
There'd be one humpty-dumpty left sitting on the wall.

Fly Away

Ten little sparrows cheeping at the door,
Six flew away and then there were four.

Ten big magpies sitting in a tree,
Seven flew away and then there were three.

8

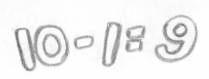

Ten shiny mynah birds sitting on the line,
One flew away and that left nine.

Ten white seagulls sitting in the sun,
Nine flew away and that left one.

The Birds Say

If you should see birds, sitting in a row,
Count them up quickly your fortune to know.

One, you'll be happy,

Two, you'll be sad,

Three, for some good news,

Four, news that's bad.

Five, you'll have silver,

Six, you'll have gold,

Seven for a secret,
That's never been told!

Doubles

One and one are two,
One for me and one for you.

Two and two are four,
That's a couple more.

Three and three are six,
Straight and crooked sticks.

Four and four are eight,
Seagulls on a gate.

11

Five and five are ten,
Tall and short men.

Six and six are twelve,
Puppies dig and delve.

Seven and seven are fourteen,
Naughty ghosts are haunting.

Eight and eight are sixteen,
Mother's busy mixing.

Nine and nine are eighteen,
Passengers are waiting.

Ten and ten are twenty,
We have all got plenty.

Skittles

Ten skittles standing straight and tall,
All lined up against the wall.
Yellow, red, green, blue and brown,
Who can come and knock some down?
Take the big ball soft and round,
Bowl it straight along the ground.
Two skittles wobble and down they fall,
How many left against the wall?

Eight skittles standing straight and tall,
All lined up against the wall.
Yellow, red, green, blue and brown,
Who can come and knock some down?
Take the big ball soft and round,
Bowl it straight along the ground.
Four skittles wobble and down they fall,
How many left against the wall?

8-4=?

Autumn

Leaves were growing on a tree,
The wind has blown some down I see.
Four have drifted far away,
Sixteen upon the branches stay.

Leaves were growing on a tree,
The wind has blown more down I see.
Eight have drifted far away,
Twelve upon the branches stay.

Leaves were growing on a tree,
The wind has blown more down I see.
Fifteen have drifted far away,
Five upon the branches stay.

Leaves were growing on a tree,
Five more have blown down I see.
Twenty leaves, the tree is bare,
Autumn's gone and winter's here.

Apples

I found a lovely apple tree,
Lots of apples are for me.
I shook the tree as hard as I could,
Down came some apples, mmm they're good.
How many apples were on that tree?
How many apples fell on me?
Seventeen apples, minus four,
I think I'll shake it just once more.

Thirteen apples on the tree,
This time eight apples fell on me.
How many left upon the tree?
Count the apples and you'll see.
The first time that I shook the tree,
Four rosy apples fell on me.
The second time I shook off eight,
So that makes twelve minus one I ate.

Stupid Animals

Oh no! The cows are out,
Eating the flowers and tramping about.
Twelve are black and four are brown,
How many cows are loose in town?

Oh no! The hens are out,
Pecking the peas and scratching about.
Fifteen white, and one that's yellow,
I'll never catch that stupid fellow!

$14=?$ $15+1=?$ $8+6=?$ $10+8=?$

Oh no! The pigs are out,
Eating the pumpkins and snorting about.
Eight run one way, six the other,
Come and help me chase them Mother.

Oh no! The sheep are out,
Eating the shrubs and wandering about.
Ten are white and eight are black,
How many sheep do I have to chase back?

21

Toucannery

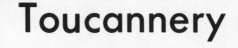

Whatever one toucan can do
Is sooner done by toucans two.

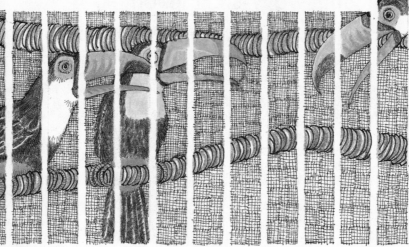

And three toucans it's very true
Can do much more than two can do.

And toucans numbering two plus two can
Manage more than all the zoo can.
In fact there is no toucan who can
Do what four or three or two can.

Age

Mary's five and Bea is three,
Bob is nine, that's three times Bea.
In four more years I'll be eleven,
That's much better than being seven.
Grandad's sixty I was told,
How many years till I'm that old?
Funny how Mum's age never changes,
She's been twenty-one for ages!